Oliver Joseph L

On Intermittent Currents and the Theory of the Induction-Balance

Elibron Classics
www.elibron.com

Elibron Classics series.

ISBN 0-543-70097-6 (paperback)
ISBN 0-543-70096-8 (hardcover)

This Elibron Classics Replica Edition is an unabridged facsimile of the edition published in 1880 in London.

From the PHILOSOPHICAL MAGAZINE for February 1880.

On Intermittent Currents and the Theory of the Induction-balance. By OLIVER J. LODGE, *D.Sc.**

1. THE telephone, considered as a scientific instrument, seems destined to play an important part as a detector of minute currents of rapidly changing intensity; and the general theory of intermittent currents is being brought into prominence by its use.

The equations to which most attention has been hitherto directed have been those relating to the steady flow of a current after the initial inductive or inertia-like effects have subsided; and in arrangements such as the Wheatstone bridge, a double key is commonly used, in order to allow the introductory stage to subside before any observation is taken. The galvanometer is essentially an instrument for measuring steady currents, or for giving the algebraically integrated expression for the total quantity of electricity which has passed in the case of transient currents; that is, $\int_0^\tau idt$, τ being small compared with the period of swing of the galvanometer-needle.

Again, the electrodynamometer has an important use as an integrator of the current without paying attention to sign;

* A portion read at the Meeting of the Physical Society on the 24th of January, 1880.

that is, its indications give the value $\frac{1}{\tau}\int_0^\tau i^2 dt$, where again τ, the total duration of the current, is small.

But the telephone-plate has such a very small period of swing that the same τ, which is vanishing compared with the time of oscillation of a needle, may be many times greater than that of the telephone-plate. Moreover the plate is not limited to one mode of vibration, but can have minor vibrations superposed on the fundamental; so that it can enter into the changes going on, and render minute fluctuations audibly apparent which would in more slowly moving detectors be simply merged in the total effect.

Thus a rapidly alternating current (such as the telephone itself produces), which is totally unfelt by a galvanometer, is appreciated by a telephone-plate in its variability—the pitch of the note indicating the number of vibrations, even if they are so rapid as to produce a shrill whistle. The telephone, in fact, does not integrate the current, but gives all its fluctuations within certain limits.

The complete theory of the telephone, setting forth precisely on what the loudness of its indications depends, would be a most interesting and important investigation; but if it has been attacked, I am ignorant of it. It seems probable that the loudness of the sounds will be found to depend upon the amplitude of the vibrations of the plate and upon their velocity conjointly—in other words, both on the total change of the current and on the rate at which the change takes place; *i. e.*, in the symbols hereafter to be used, that the loudness is a direct function of $j\frac{dj}{dt}$. I shall not, however, assume anything of this sort, but shall content myself with simply finding the value of the current j as a function of the time, leaving the rest to be done subsequently.

It is quite true that the telephone is only an indicator and not a measuring-instrument; but so many null methods can be devised which permit measurements to be made with a simple detector, that it is probable that it will have important applications in this capacity also. And hence I think the general theory of intermittent currents, or of currents in the variable stage, will come into more prominence than hitherto.

The induction-balance furnishes an illustration. Dove made experiments with it; and Felici established the laws of current induction by its aid: but its power as an instrument of research was never appreciated till Prof. Hughes applied to it an intermittent current and a telephone.

Faraday interposed blocks of copper between a primary coil and a secondary connected with a galvanometer, and was surprised to find that the effect at make and break was precisely the same with the copper as without it*. It was, however, afterwards found that the physiological effects were very different, being much less when the copper was present—thus proving that though the copper did not affect the integral flow of electricity, yet it greatly affected the time during which that flow took place.

Dove repeated Faraday's experiment more perfectly by means of an induction-balance, and showed that no non-magnetic media produced any effect appreciable by a galvanometer.

Thus may numerous phenomena be overlooked with a galvanometer which with a telephone become obtrusively evident.

2. My attention was more particularly directed to the subject by an observation which was made by Mr. W. Grant, assistant in the Physical Laboratory, University College, and which I have his permission to describe.

The intermittent current from a mouth-telephone, or the induced current from a clock-ticking microphone and coil, is sent through a long helix of wire wound upon a bobbin, with another similar but quite disconnected wire wound alongside it. A telephone and ear are also arranged in the circuit of the first wire, and the loudness of the sound observed. The disconnected wire, which is wound on the same reel as the first, now has its two ends joined up so that it itself forms a closed circuit: the loudness of the sound is thereby considerably increased. The secondary circuit is broken again, and the sound again becomes faint. The strengthening on closing the secondary circuit is so great, indeed, that short-circuiting the primary so as to shunt out all its resistance, scarcely produces any additional strengthening effect. In other words, a moderate resistance (several ohms) may be thrown into a telephonic circuit without the slightest appreciable weakening of the effect, provided a second wire coiled alongside the first be arranged so as to form a closed circuit. If a second telephone be put in circuit with this second wire, it will give about the same indications as the other; and this may sometimes be a good way of working two telephones.

The rough general explanation is, of course, not far to seek. The extra currents, which at first oppose the primary, are able to form in the secondary when its circuit is closed; and hence the changes in strength of the primary are more rapid, and therefore more complete.

* Exp. Res. arts. 1721, 1735.

An obvious modification is, to arrange both speaking and hearing telephone in circuit with both the parallel coiled wires in series, first, so that the currents flow the same way in both wires, and then with one wire reversed; so that in the first case the self-induction is a maximum, and in the other almost zero.

Any one who tries this experiment will at once appreciate the very decided difference in the telephonic indications. In fact, resistance is of much less consequence in a telephone-circuit than is self-induction. This fact obviously explains the discrepancy which has been observed between experiments on resistance-bobbins thrown into a telephonic circuit in a laboratory, and the same resistance when forming an actual line. The resistance-coils are wound so as to have no self-induction; and they accordingly weaken the effect very little. It is surprising how many thousand ohms can be thrown into a talking circuit without stopping a conversation. But if the bobbins were wound all in one direction, they would be still worse than a line-resistance.

3. It is plain from this that the only right method of arranging a telephone-line is to have two wires running close together—one for the direct, the other for the return current. This arrangement has, I believe, been used in order to avoid induction-effects from battery-currents in neighbouring wires; but it has also the great advantage of diminishing the self-induction of the wire itself. It is, however, a question how far the increase in electrostatic capacity would militate against this advantage in a very long land-line. It is certain to be injurious as far as it goes; but if the wires are very thin, their electrostatic capacity will be but small.

In the case of a cable the double wire would have decidedly less capacity than a single wire; and the capacity diminishes rapidly with the size of the wire. Hence it appears that a telephone-cable should have a pair of exceedingly fine insulated wires of the very highest conductivity pretty close together in a large insulating sheath.

I need hardly say, perhaps, that no advantage will be gained by attempting to do away with the self-induction of the telephone-bobbin itself by means of a second wire wound alongside the first and forming a closed circuit—because the intermittent currents induced in this would be nearly as strong as those in the direct circuit, and, being always in an opposite direction, will nearly neutralize their intermittent effect on external coils or on the telephone-plate.

I propose now to work out the theory of the induction-balance, and of one or two other cases of intermittent currents, as completely as possible without taking into account the electrostatic capacity of the wires. It would be very interesting to obtain the perfectly general equations; but they are very complex, and appear to be beyond me at present.

General Theory of Current-induction.

4. The general theory of the establishment of a current in a number of circuits of known resistance, in one or all of which a given electromotive force begins to act, was begun by Helmholtz and Thomson, who independently showed that the fact of the existence of induced currents, discovered experimentally by Faraday, as well as the laws of the same, could be deduced mathematically from the electromagnetic attractions of Oersted and Ampère by means of the principle of the conservation of energy. The complete theory, it is needless to say, is to be found in Maxwell's 'Electricity,' and also in that admirable digest of the whole subject, Professor Chrystal's article in the *Encyclopædia Britannica.*

Suppose there are n circuits of resistances $r_1, r_2, \ldots,$ in each of which "impressed" electromotive forces $E_1, E_2 \ldots$ begin to act at the same instant; then the strengths of the currents $i_1, i_2 \ldots$ in the several circuits at any time t afterwards are to be found by the following set of n equations—

$$\left.\begin{aligned} E_1 - r_1 i_1 &= \frac{d}{dt}(M_{11}i_1 + M_{12}i_2 + \ldots + M_{1n}i_n), \\ E_2 - r_2 i_2 &= \frac{d}{dt}(M_{21}i_1 + M_{22}i_2 + \ldots + M_{2n}i_n), \\ &\cdot\quad\cdot\quad\cdot\quad\cdot\quad\cdot\quad\cdot\quad\cdot\quad\cdot \\ E_n - r_n i_n &= \frac{d}{dt}(M_{n1}i_1 + M_{n2}i_2 + \ldots + M_{nn}i_n), \end{aligned}\right\} \quad \ldots \quad (1)$$

where M_{lk} is a double integral, called the coefficient of mutual induction between the circuit l and the circuit k.

Induction-balance.

5. Now in the induction-balance as ordinarily arranged there are three such circuits, with the coefficients M all constant, and two of them without any electromotive force. The first, or battery-circuit, consists of a battery and two equal coils, and it also contains some kind of intermittent break or microphone; the second is the telephone-circuit, and consists of two equal coils and a telephone; and the third is complete within the coin or other piece of conducting matter in the neighbourhood of one pair of the coils of the other circuits. The general value of the telephonic current at any instant after a make has occurred in the primary would therefore be

obtained by the solution of a linear differential equation of the third order with constant coefficients.

But the problem may be simplified by observing that the "coin" is not introduced until a perfect "balance" has been obtained between the first and second circuits—that is, until they are conjugate to one another so that $M_{12}=0$. Moreover the coefficients of induction between the two coiled circuits and the coin are ordinarily sufficiently small for their squares to be neglected; in which case, as will be shown more fully later (sections 12 and 13), the above set of equations reduces to the following—

$$M_{11}\frac{di_1}{dt}+ \quad 0 + \quad 0 + r_1 i_1 = E,$$

$$0 + M_{22}\frac{di_2}{dt} + M_{23}\frac{di_3}{dt} + r_2 i_2 = 0,$$

$$M_{31}\frac{di_1}{dt} + \quad 0 + M_{33}\frac{di_3}{dt} + r_3 i_3 = 0,$$

or, as I shall for greater convenience write them,

$$\left.\begin{aligned} &L\frac{di}{dt}+Ri=E, \\ &l\frac{dj}{dt}+m\frac{dk}{dt}+rj=0, \\ &\lambda\frac{dk}{dt}+\mu\frac{di}{dt}+\rho k=0, \end{aligned}\right\} \quad . \quad . \quad . \quad . \quad (2)$$

where the three coefficients of self-induction are written L, l, and λ (λ being very small); and where the battery-current is i the telephone-current is j and is small, and the current in the coin is k; m signifies the effect of the coin-current on the telephone-circuit, and μ the effect of the primary circuit on the coin. The effect of the primary on the secondary (M) is, as already explained, supposed to be 0; that is, balance is supposed to have been obtained before the insertion of the coin. ρ, or the resistance which the current circulating in the coin experiences, is a quantity which is likely to be several times larger than the specific resistance of the material (*i. e.* the resistance of a centimetre cube); but for ordinary coins it is a number of the same order of magnitude. It must be remembered that R and L stand for the resistance and self-induction of the *whole* of the battery-circuit, and, similarly, r and l include the telephone-coil as well as the other two coils of the telephone-circuit.

Now the only difficulty in solving these equations consists in the varying resistance of the battery-circuit, in which a contact-breaker, or clock-ticking microphone, or some other arrangement for producing an intermittent current is inserted. I shall therefore leave the consideration of the induction-balance for the present, and examine the case of a primary circuit in space by itself and having a break and make in some part of it. Then we will consider the case of an intermittent primary in the neighbourhood of a closed secondary; and after this it will be easy to apply our results to the induction-balance.

On the Law of Variation of a Battery-current in a solitary Circuit.

6. Let a battery of constant electromotive force E act in a circuit whose total resistance may be *suddenly* changed from the value R to the value S, and let L be the coefficient of self-induction of this circuit. Then the current i at any time t after the change of resistance has occurred is to be found from the equation

$$L\frac{di}{dt}+Si=E, \quad . \quad . \quad . \quad . \quad . \quad . \quad . \quad (3)$$

with the initial condition $i=\frac{E}{R}$ when $t=0$. This gives us in the integral form

$$i=\frac{E}{S}\left(1+\frac{S-R}{R}e^{-\frac{S}{L}t}\right). \quad . \quad . \quad . \quad . \quad . \quad (4)$$

Putting $R=\infty$, we get the well-known expression for the current at "make,"

$$i=\frac{E}{S}(1-e^{-\frac{S}{L}t}). \quad . \quad . \quad . \quad . \quad . \quad . \quad (5)$$

Putting S very large, we get an expression for the current at partial break, provided there is no extra-current spark at the surface of separation,

$$i=\frac{E}{R}e^{-\frac{S}{L}t}. \quad . \quad . \quad . \quad . \quad . \quad . \quad . \quad . \quad (6)$$

It will not do to put $S=\infty$, because it is impossible to make the resistance *suddenly* infinite, *i. e.* to stop the current instantaneously.

7. It is difficult to get a reasonably correct expression for the value of the current at break. We may suppose that, by separation of two portions of the circuit, the resistance is suddenly changed from R to a quantity which would be S if the current instantaneously ceased, but which is much less than S for a very short time, owing to the heat generated by the

current itself. The temperature of the spark at any time t after the break is to be found, we may suppose, from such an equation as this,

$$mc\theta = \int_0^t (ri^2 - H\theta)dt,$$

where H is the cooling-constant. The mode in which the resistance of air changes with temperature is unknown; so we may assume any simple and not improbable law, such as that the decrement of resistance is proportional to the increment of temperature, or

$$dr = -k\,d\theta,$$

where k is a constant. From these two equations r has to be obtained as a function of the current i and of the time t; and then its value has to be substituted for S in equation (3). As a first approximation, we may imagine the time too short for cooling, or make H=0. Then we get

$$r = Se^{-k'\int_0^t i^2 dt};$$

an expression which, as a coefficient in equation (3), to me seems quite unmanageable. I will therefore assume for the future that the change in the battery-circuit resistance is always made suddenly from one finite value to another, no spark or current across air-spaces being produced.

8. *Suppose the resistance of a battery-circuit oscillates rapidly between the values* R *and* S, *each change being made suddenly and lasting for the short time* τ; *what is the strength of the current after a few seconds?*

This may be assumed to be the sort of thing that happens in a microphone or other arrangement for producing an intermittent current.

The strength of the current before the vibration begins is

$$i_0 = \frac{E}{R};$$

at the end of the first period τ the strength is, by equation (4),

$$i_1 = \frac{E}{S} + \frac{E(S-R)}{SR} e^{-\frac{S}{L}t};$$

at the end of the second period τ the current is

$$i_2 = \frac{E}{R} - \frac{E(S-R)}{SR} \left(e^{-\frac{R}{L}\tau} - e^{-\frac{R+S}{L}\tau}\right),$$

and so on—

$$i_4 \text{ being } \frac{E}{R} - \frac{E(S-R)}{SR}(\text{exp.})\{(R)-(R+S)+(2R+S)-(2R+2S)\}\left(-\frac{\tau}{L}\right)$$

and

$$i_5 \text{ being } \frac{E}{S} + \frac{E(S-R)}{SR}(\text{exp.})\{(S)-(S+R)+(2S+R)-(2S+2R)+(3S+2R)\}\left(-\frac{\tau}{L}\right)$$

Accordingly, after a very large number of vibrations (which will be accomplished in a few seconds) the current is either

$$i = \frac{E}{S} + \frac{E(S-R)}{SR}\left(\frac{1-e^{-\frac{R}{L}\tau}}{1-e^{-\frac{S+R}{L}\tau}}\right)e^{-\frac{S}{L}t}, \quad . \quad . \quad (7)$$

or else the same expression with R and S everywhere interchanged, according as the last change of resistance was from R to S or from S to R, t being the time which has elapsed since the last change.

The law of variation of such a current is therefore just the same as for a simple make and break (4); but the part depending on time is multiplied by a constant fraction always less than unity, and which diminishes rapidly as τ (the period of a semi-vibration of the intermittence) diminishes.

If τ be infinitesimal compared with $\frac{L}{R+S}$, the fraction is $\frac{R}{S+R}$, and hence the current is

$$i = \frac{E}{S}\left(1 + \frac{S-R}{S+R}e^{-\frac{S}{L}t}\right),$$

or, as of course t (being less than τ) is itself vanishing,

$$i = \frac{2E}{S+R}. \quad . \quad . \quad . \quad . \quad . \quad . \quad (8)$$

Hence the current tends to approach this constant value if the vibrations are too rapid. Such a current as this is of course inappreciable by a telephone; but the above would be its galvanometric indication.

This must be something like the state of things in a coil with a too rapid break.

Query, whether any thing of the same sort happens when a battery-current is passed through a vacuum-tube, the intermittence being almost too rapid to be heard by the telephone. Drs. De la Rue and Hugo Müller have put a telephone in circuit with a battery and vacuum-tube, and heard only a faint rustling when the stratifications were steady.

Equation to a continuously Intermittent Current.

9. It would be interesting to obtain an expression for the current when the resistance of the circuit varies *continuously* from some finite value to infinity and back again in a short regular period $\frac{\pi}{\omega}$. Thus R might be supposed to be $\rho \sec^2 \omega t$; and the equation would be

$$L\frac{di}{dt}+\rho i \sec^2 \omega t = E, \quad . \quad . \quad . \quad . \quad . \quad . \quad . \quad (9)$$

of which the solution is

$$i = Ce^{-\frac{\rho}{L}\tan \omega t} + \frac{E}{L}e^{+\frac{\rho}{L}\tan \omega t}\int e^{-\frac{\rho}{L}\tan \omega t}dt; \quad . \quad . \quad (10)$$

but the integral $\int e^{\tan x}dx$ does not appear to be evaluable except by an unmanageable lot of series.

Equation to an Alternating Current.

10. Instead of varying the resistance, we may produce an alternating current by a periodic electromotive force, as, for instance, when the circuit contains an electromagnetic machine of constant resistance, such as a telephone itself. The electromotive force is then representable by a simple harmonic function or by a sum of a number of these. Take the simplest case,

$$L\frac{di}{dt}+Ri = E \sin 2\pi n t, \quad . \quad . \quad . \quad . \quad (11)$$

where the electromotive force oscillates from E to $-$E and back again n times a second. The solution of this is

$$i = Ce^{-\frac{R}{L}t} + \frac{\frac{R}{L}\sin 2\pi n t - 2\pi n \cos 2\pi n t}{\frac{R^2}{L^2}+4\pi^2 n^2}\cdot\frac{E}{L}. \quad . \quad (12)$$

The first term rapidly dies out; and so the permanent value is

$$i = \frac{E}{R}\cdot\frac{\sin \omega t - \lambda\omega \cos \omega t}{1+\lambda^2\omega^2}, \quad . \quad . \quad . \quad . \quad . \quad . \quad . \quad . \quad (13)$$

writing ω for $2\pi n$ and λ for the time-constant $\frac{L}{R}$.

On the Law of Variation of a Battery-current in a Polarizable Circuit.

11. In the previous sections we have considered the electromotive force E of the battery to be constant. Now this

is never accurately true, as the electromotive force begins to fall off the instant the circuit is closed, and some discrepancies between theory and experiment may arise from this cause. Let us see what happens when the battery is not constant, or when a polarizable voltameter forms part of the circuit. According to Kohlrausch, the electromotive force of polarization is proportional to the amount of decomposition, and therefore a short time t after the current has been established it is

$$p\int_0^t idt,$$

where p is the electromotive force produced by the passage of a unit of electricity. The equation to the current is therefore

$$L\frac{di}{dt}+Ri=E-p\int_0^t idt,$$

or

$$L\frac{d^2i}{dt^2}+R\frac{di}{dt}+pi=0\,; \quad . \quad . \quad . \quad . \quad . \quad (14)$$

and the integral of this, remembering that $i=0$ when $t=0$, is

$$i=\frac{2E}{R}e^{-\alpha t}\sinh\beta t, \quad . \quad . \quad . \quad . \quad . \quad (15)$$

where

$$\alpha=\frac{R}{2L} \text{ and } \beta=\sqrt{\left(\frac{R^2}{4L^2}-\frac{p}{L}\right)}.$$

If $p=\frac{R^2}{4L}$, there could never be any current; and it is impossible for p to be greater than $\frac{R^2}{4L}$. Practically, however, p would always be very small compared with this quantity; and so we may write

$$\beta \doteq \frac{R}{2L}-\frac{p}{R}.$$

Hence in any ordinary polarizable circuit the strength of the current at any time after "make" is

$$i=\frac{E}{R}\left(e^{-\frac{p}{R}t}-e^{\left(\frac{p}{R}-\frac{R}{L}\right)t}\right). \quad . \quad . \quad . \quad . \quad (16)$$

On the Law of Variation of a Battery-current when a closed secondary circuit is stationary in its neighbourhood.

12. So far we have considered a primary circuit in space by itself; but now we will arrange near it a secondary coil with a resistance r and a coefficient of self-induction l, and

we will suppose M to be the coefficient of mutual induction between the primary and secondary. The coils shall be fixed in position so that M is constant; then a current j will be developed in the secondary circuit whenever the primary changes from R to S; and this will react on the primary current i, the equations for determining both being

$$L\frac{di}{dt}+M\frac{dj}{dt}+Si=E \quad . \quad . \quad . \quad . \quad (17)$$

and

$$l\frac{dj}{dt}+M\frac{di}{dt}+rj=0, \quad . \quad . \quad . \quad . \quad (18)$$

with the conditions $i=\frac{E}{R}$ and $j=0$ when $t=0$.

Differentiating these (E being supposed constant), we easily separate the variables and obtain

$$(Ll-M^2)\frac{d^2i}{dt^2}+(Lr+Sl)\frac{di}{dt}+Sri=E, \quad . \quad . \quad (19)$$

and

$$(Ll-M^2)\frac{d^2j}{dt^2}+(Lr+Sl)\frac{dj}{dt}+Srj=0. \quad . \quad . \quad (20)$$

Now the solution of (20), remembering the initial condition, is

$$j=Je^{-\alpha t}\sinh\beta t \quad . \quad . \quad . \quad . \quad . \quad . \quad . \quad . \quad . \quad . \quad (21)$$

or

$$=\tfrac{1}{2}J\{e^{-(\alpha-\beta)t}-e^{-(\alpha+\beta)t}\},$$

where

$$\alpha=\frac{Lr+lS}{2(Ll-M^2)} \text{ and } \beta=\frac{\sqrt{\{(Lr-lS)^2+4rSM^2}}{2(Ll-M^2)} \quad . \quad (22)$$

$$\left[\text{i. e. } \alpha^2-\beta^2=\frac{rS}{Ll-M^2}\right],$$

and where J is a constant which has yet to be determined. It is well to notice that neither α nor β can vanish or become imaginary; but they can become infinite simultaneously.

To obtain the value of the primary current i, we can either write down the solution of (19), viz.

$$i=\frac{E}{S}+e^{-\alpha t}(C_1e^{\beta t}+C_2e^{-\beta t}); \quad . \quad . \quad . \quad . \quad (23)$$

or, what is practically more convenient as determining all arbitrary constants directly, we can combine equations (17) and

(18) so as to eliminate $\frac{di}{dt}$, and then put in the values of j and $\frac{dj}{dt}$ from (21). We thus get

$$M(Si-E)=\tfrac{1}{2}Je^{-\alpha t}\{\sqrt{(Lr-Sl)^2+4rSM^2}.\cosh\beta t + (Lr-lS)\sinh\beta t\},$$

which will give us the value of J, since $i=\frac{E}{R}$ when $t=0$. We find then that

$$J=\frac{2ME(S-R)}{R\sqrt{\{(Lr-lS)^2+4rSM^2\}}}; \quad . \quad . \quad . \quad . \quad (24)$$

and this is the value to be substituted in equation (21).

Hence, finally, we can write out explicitly the values of i and j, though, as the constants are very long, we will make an abbreviation by writing

$$\frac{4rSM^2}{(Lr-lS)^2}=\lambda^2; \quad . \quad . \quad . \quad . \quad . \quad . \quad . \quad (25)$$

then the strength of the battery-current at any time after the resistance of its circuit has suddenly jumped from R to S is

$$i=\frac{E}{S}+\frac{E(S-R)}{SR}e^{-\alpha t}\left\{\cosh\beta t+\frac{\sinh\beta t}{\sqrt{(1+\lambda^2)}}\right\}; \quad . \quad (26)$$

and the strength of the induced current at the same instant is

$$j=\frac{2ME(S-R)}{R(Lr-lS)\sqrt{(1+\lambda^2)}}e^{-\alpha t}\sinh\beta t. \quad . \quad . \quad . \quad (27)$$

We may notice that the expression for i contains only the *square* of M; that is, the square of M expresses the reaction of the secondary on the primary; hence when M is small this may be neglected.

Special Cases.

13. It will be interesting now to consider the special cases in which these equations (26) and (27) may be expected to assume simple forms.

Case 1. When the coils are so far apart that M^2 is negligible compared with Ll.

In this case $\lambda=0$, $\alpha+\beta=\frac{r}{l}$, $\alpha-\beta=\frac{S}{L}$; and it will be found that the expression for the primary current i as given in equation (26) reduces to the expression (4), which we found for the current in a solitary primary circuit, as it evidently ought to do; and there is nothing new to be said about it.

But the value of j, the current induced in the secondary circuit, becomes

$$j=\frac{\mathrm{ME(S-R)}}{\mathrm{R(L}r-l\mathrm{S)}}\left(e^{-\frac{\mathrm{S}}{\mathrm{L}}t}-e^{-\frac{r}{l}t}\right). \quad . \quad . \quad . \quad (28)$$

To find the induced current at instantaneous break in the special case now being considered, we must put $\mathrm{S}=\infty$, and we get

$$j=\frac{\mathrm{ME}}{l\mathrm{R}}e^{-\frac{r}{l}t}, \quad . \quad . \quad . \quad . \quad . \quad . \quad . \quad (29)$$

which is the recognized value for it (see Chrystal, 'Encyc. Brit.' equation (41)). Its initial or maximum value is $\frac{\mathrm{M}}{l}\cdot\frac{\mathrm{E}}{\mathrm{R}}$; and it rapidly dies away.

To find the induced current at "make" we must put $\mathrm{R}=\infty$; and then (28) becomes

$$j=-\frac{\mathrm{ME}}{(\mathrm{L}r-l\mathrm{S})}\left(e^{-\frac{\mathrm{S}}{\mathrm{L}}t}-e^{-\frac{r}{l}t}\right). \quad . \quad . \quad . \quad (30)$$

This, therefore, begins at zero, rises to a maximum after the lapse of time

$$t=\frac{\mathrm{L}l}{\mathrm{L}r-l\mathrm{S}}\log\frac{\mathrm{L}r}{l\mathrm{S}},$$

and then dies away.

If the primary and secondary circuits are *similar*, so that

$$\frac{\mathrm{S}}{\mathrm{L}}=\frac{r}{l},$$

the expression for the induced current at make simplifies, becoming

$$j=-\frac{\mathrm{ME}}{\mathrm{L}l}te^{-\frac{r}{l}t}. \quad . \quad . \quad . \quad . \quad . \quad . \quad (31)$$

14. *Case* 2.—When the primary and secondary circuits are wound side by side, so that the coefficient of mutual induction nearly equals the coefficient of self-induction of either—in other words, so that $\mathrm{L}=l=\mathrm{M}$.

For this case (see equations 22 and 25),

$$\lambda=\frac{4r\mathrm{S}}{(r-\mathrm{S})^2}, \quad \text{and } \alpha=\beta=\infty;$$

but if we put $\mathrm{L}-\mathrm{M}=$ a small quantity z, say, and then proceed to the limit, we shall find a finite value for the difference of α and β, viz.

$$\alpha-\beta=\frac{rS}{L(r+S)},$$

while

$$\alpha+\beta=\frac{r+S}{2z};$$

and accordingly we get as the value of the primary current (26),

$$i=\frac{E}{S}\left\{1+\frac{S-R}{R(r+S)}\left\{re^{-\frac{rS}{L(r+S)}t}+Se^{-\frac{r+S}{2z}t}\right\}\right\}. \quad (32)$$

The last term in these brackets decreases at a nearly infinite rate; hence the primary current jumps almost suddenly from the value $\frac{E}{R}$ to the value

$$\frac{E}{R}\cdot\frac{r+R}{r+S}, \quad \ldots \ldots \quad (33)$$

and then increases (or decreases, as the case may be) at a more moderate pace, its subsequent values being given by the equation

$$i=\frac{E}{S}\left(1+\frac{r(S-R)}{R(r+S)}e^{-\frac{rS}{L(r+S)}t}\right). \quad \ldots \quad (34)$$

The form of expression (32) when S is made very great is noticeable, as it shows that the primary current in a wire coiled up with a closed secondary is able to stop nearly as dead as if the primary were doubled upon itself (6). It is

$$i=\frac{E}{R}e^{-\frac{S}{2z}t}. \quad \ldots \ldots \quad (35)$$

The value of the induced current under the same circumstances (viz. when the two circuits are coiled close together) is easily obtained from (27), and is

$$j=\frac{E(S-R)}{R(r+S)}e^{-\frac{rS}{l(r+S)}t}. \quad \ldots \ldots \quad (36)$$

If $S=\infty$, this is the current at instantaneous break, viz.

$$j=\frac{E}{R}e^{-\frac{r}{l}t}, \quad \ldots \ldots \quad (37)$$

the same as (29) would have given; but if $R=\infty$, it is the current at make, viz.

$$j=-\frac{E}{r+S}e^{-\frac{S}{r+S}\cdot\frac{r}{l}t}. \quad \ldots \ldots \quad (38)$$

The induced current at *make* in this case therefore has an instantaneous maximum value $\frac{E}{r+}$ and it then dies away at about half the rate of the current at break.

All this is very instructive. It must be the state of things approximated to in many arrangements.

15. We can now see what happens in Mr. Grant's case of a primary wound alongside a secondary, whose circuit can be closed or unclosed; though the state of things considered in the last section is only roughly attained, because the telephone-coil forms part of the battery-circuit, and adds materially to the self-induction of that circuit without contributing to the mutual induction. However, neglecting this and taking the circuits similar, we find the primary current at "make" ($R=\infty$, $S=r$), by equation (32), to be

$$\frac{E}{S}\left(1-\tfrac{1}{2}e^{-\frac{S}{2L}t}-\tfrac{1}{2}e^{-\frac{S}{z}t}\right),$$

whereas if the secondary circuit had been unclosed it would have been (5)

$$\frac{E}{S}\left(1-e^{-\frac{S}{L}t}\right).$$

Similarly, at the break ($S=$ large, $R=r$), the primary current, when the secondary is closed, is

$$\frac{E}{R}e^{-\frac{S}{2z}t}$$

instead of

$$\frac{E}{R}e^{-\frac{S}{L}t}$$

when it was open.

Hence the initial or maximum rate of variation of the battery-current either at make or break when the secondary current is closed, is to its initial rate of variation when the secondary is unclosed as L is to $2(L-M)$,—a ratio which, in the case supposed, is nearly infinite.

The Theory of Tertiary Currents as produced in an induction-balance—that is, in a coil which is arranged so as to be conjugate to the primary coil, but which is in the neighbourhood of a third coil or conductor affected inductively by the primary current.

16. This is the case of the induction-balance disturbed by a coin. The primary and secondary circuits are arranged in some conjugate position so that they have no effect on each

other—this being done practically by splitting each into two equal parts, and arranging the two parts so as to oppose each other; and the third circuit is represented by the piece of metal put near or inside the primary and secondary coils. The primary circuit contains as part of itself a battery and an intermitter; the secondary contains a telephone.

Neglecting electrostatic capacity and leakage (which latter, by the way, must be carefully avoided in a good balance), the current in either primary is the same, and the current in each secondary is the same, not only on the whole, but at every instant of time. In fact, the separation of the two halves of the primary and secondary is immaterial to the theory. The current which is induced in the secondary circuit, and which is heard in the telephone, is a tertiary current produced by induction from the third coil or piece of conducting matter.

The currents in the battery-circuit at any instant being called i, that induced in the metal being k, and that induced by this in the secondary circuit being called j, and being the value appreciated by the telephone, the equations for determining these three quantities are given in § 5, on the supposition that the coefficients of mutual induction between the metal and the two coils respectively (m and μ) are small numbers, which is always true in practice.

The battery-circuit is then practically undisturbed by the presence of the other parts of the instrument, and may be regarded as in space by itself (see end of §§ 12 and 13); so that all we have said about the intermittence of currents in such a circuit in sects. 6–11 applies to the primary of an induction-balance: we will, however, only take the case when the resistance jumps suddenly from any finite value R to another finite value S. The value of i is then given by equation (4); and its rate of variation is

$$\frac{di}{dt}=-\frac{\mathrm{E(S-R)}}{\mathrm{RL}}e^{-\frac{\mathrm{S}}{\mathrm{L}}t};$$

but as (equation 2)

$$\lambda\frac{dk}{dt}+\rho k+\mu\frac{di}{dt}=0,$$

it is easy to determine k; viz.

$$k=\frac{\mu\mathrm{E(S-R)}}{\mathrm{R(L}\rho-\mathrm{S}\lambda)}\left\{e^{-\frac{\mathrm{S}}{\mathrm{L}}t}-e^{-\frac{\rho}{\lambda}t}\right\}. \quad . \quad . \qquad (39)$$

This value agrees exactly with that obtained in the general case (§ 12) when the condition that M^2 might be neglected was put in (see eq. 28), and therefore confirms the correctness of our working. All that was said about the special cases of j in

§ 13 is therefore now true of k. What we want to find, however, is not k but j, the current induced in the telephonic circuit; hence, differentiating the above expression for k and substituting it in

$$l\frac{dj}{dt}+rj+m\frac{dk}{dt}=0,$$

we get, remembering the initial condition that $j=0$ when $t=0$,

$$j=-\frac{m\mu E(S-R)}{R}\left\{\frac{r}{(Lr-lS)(l\rho-\lambda r)}e^{-\frac{r}{l}t}\right.$$

$$\left.+\frac{\rho}{(l\rho-\lambda r)(\lambda S-L\rho)}e^{-\frac{\rho}{\lambda}t}+\frac{S}{(\lambda S-L\rho)(Lr-lS)}e^{-\frac{S}{L}t}\right\}. \quad (40)$$

And this is the current heard in the telephone.

The expression $\int_0^\infty jdt$ identically vanishes; hence this current cannot affect a galvanometer (see § 1).

At "make" $R=\infty$, and the three terms in the brackets are all present, the factor outside reducing to $+m\mu E$. But at "break" $S=\infty$, and the third term disappears, leaving

$$j=\frac{m\mu E}{R(l\rho-\lambda r)}\left\{\frac{r}{l}e^{-\frac{r}{l}t}-\frac{\rho}{\lambda}e^{-\frac{\rho}{\lambda}t}\right\}; \quad . \quad . \quad (41)$$

which is the tertiary current at break, and has some of the characteristics of the secondary current at make, see equation (28).

17. If the battery and telephone circuits are *similar*, so that $\frac{S}{L}=\frac{r}{l}$, the value of j at the make simplifies considerably, becoming

$$\frac{m\mu ES}{Ll(\lambda S-L\rho)}\cdot te^{-\frac{r}{l}t}\Big\}, \quad . \quad . \quad . \quad . \quad (42)$$

which in this case therefore bears a constant ratio to the secondary current at make under the same circumstances (see equation 31).

Theory of the Induction-balance continued.

18. Having now integrated equations (2) and obtained an expression (40) for the current passing through the telephone at any instant, several things may be noticed about it. First, the current produced by the piece of metal is in general different in character to that produced by a slight shift of the secondary coil (that is, by M being made not quite zero); and consequently it is impossible in general to balance the effect

of metal completely by moving the other pair of coils a little further apart, though it may be done partially. Both effects would then be superposed, and the two sounds would be separately heard and might differ in pitch, because there would be more oscillations in the tertiary current than in the secondary. It is quite possible that a musical ear would perceive that the quasi-note produced by the insertion of a piece of metal contains a trace of a tone an octave above that produced by a shift of the coils. Equation (42), however, appears to show that when the battery- and telephone-circuits are *similar* the secondary and tertiary currents are expressible by the same function of the time, and that, therefore, in this case the effect of introducing metal *may* be counteracted by a shifting of the coils.

Next, we see that the effect is independent of M (that is, of the mutual induction between the primary and secondary coils); and therefore it may occasionally be well to place them in some nearly conjugate position, so as more easily to obtain a balance. In fact the four coils are unnecessary; two coils placed so as not to act on each other will do; and a piece of metal placed near them will produce disturbance; only it will not be so easy to bring the metal close or parallel to both coils.

Next the effect is proportional to the produced $m\mu$; of which μ is the potential of the primary, and m that of the secondary, on the coin. Hence, if the two coils are equal in size, every thing is symmetrical with respect to them, so that it does not matter whether the coin is nearest the primary or the secondary; moreover the middle point halfway between the two coils must be either a maximum or a minimum position of the coin.

Prof. Hughes long ago stated, from experimental observation, that the position of the coin which gave the loudest sound was half way between primary and secondary, the two coils being pretty close together and being equal in all respects. Before investigation I had felt inclined to doubt whether this were more than approximately true (as it was by no means evident *à priori* that every thing was symmetrical with respect to both coils, and the middle of the primary coil seemed a not improbable place for the maximum). Accordingly I arranged two pairs of coils as an induction-balance, the components of each pair being 4 or 5 inches apart, but with their planes parallel as usual; and I then moved a coin about along the common axis of one pair to see what happened. The maximum was *not* in the middle; but there were two maxima, one on each side the middle, and every thing appeared symmetrical with respect to both coils. The preceding investigation throws light upon this, and reconciles the two observations. It turns out (as might have been anticipated) that Professor Hughes's statement is quite correct when the coils are near together; but if they are separated by a distance greater than the diameter of either (the coin being supposed small), the middle point becomes a *minimum* with a maximum on each side of it (see next section).

On the Law according to which the disturbance produced by a small coin on the common axis of a pair of coils in an induction-balance depends on its position.

19. The general expression for the induction-coefficient between two circular coils in any position whatever is given in 'Maxwell,' art. 696. We can specialize this to the case required, viz. a coil of n turns and a "coin" or coil of one turn, both on the same axis. Let the mean radius of the "coil" be a, and let the distance of its circumference from some point O on the axis be c. Also let the mean radius of the "coin" be b, and let d be the distance of its centre from the point O. Then the mutual induction-coefficient between the coil and the coin is (art. 699)

$$2n\pi^2\frac{a^2b^2}{c^3}\left\{1+3\frac{\sqrt{(c^2-a^2)}}{c^2}d+6\frac{c^2-\frac{5}{4}a^2}{c^4}(d^2-\tfrac{1}{4}b^2)+\&c.\right\}.$$

Now d we can make zero at once by taking the point O at the middle of the coin, so that c is the distance from the mean circumference of the coil to the middle of the coin; and there remain inside the brackets of the above expression 1 + terms involving the square and higher powers of $b:c$, which is a small quantity. Thus a very good approximation to the induction-coefficient between the primary and a small coin is

$$\mu=\frac{2\pi^2a^2b^2n}{c^3}. \quad . \quad . \quad . \quad . \quad . \quad . \quad (43)$$

Similarly the induction-coefficient between the secondary and the coin is the same expression with dashed letters; hence the product

$$m\mu=\frac{nn'}{(cc')^3}(2\pi^2b^2aa')^2. \quad . \quad . \quad . \quad . \quad . \quad . \quad (44)$$

And this is a maximum when cc' is a minimum, a thing which is very easily represented geometrically.

For let A and A′ be points on the mean circumferences of the coils from which c and c' are measured; draw the system of lemniscates cc' = various constants, and then draw through the system any straight line at a distance a from A and a' from A′.

This line will represent the common axis of the two coils, and points on it represent possible positions of the coin. Now the line cuts every lemniscate in four points: hence there are four points on the axis at which the coin produces the same disturbance. These can run together two and two at points where the line is touched by the lemniscate; and these are the maximum or minimum points. They are maxima if the line is touched on the side of A A′, but minima if the line is touched on the other side.

If the coils are the same size (*i. e.* if $a=a'$), and if they are at a proper distance apart (viz. $=2a$), all four points can run together in the middle, which is therefore then a stationary point, so that moving the coin a little either way will produce very little diference. But if the coils are closer together than this, two of the points of section become imaginary, and hence there are now only two positions of the coin which give the same noise; and these two can run together in the middle, which is now a maximum.

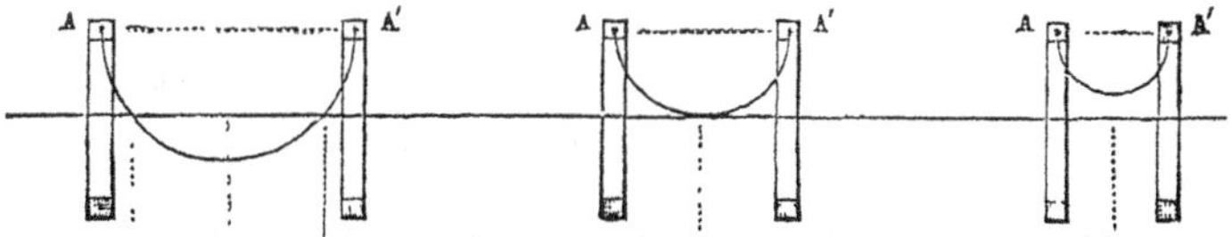

Max. Min. Max. Stationary. Max.

Analytically the problem might be stated thus:—Find the max. and min. values of $\sin\theta\sin\theta'$, given the condition $a\cot\theta+a'\cot\theta'=$ const., θ and θ' being the angles subtended by mean radii of the primary and secondary coils at the coin. If the coils are equal ($a=a'$), the solutions are $\theta=\theta'$ and $\theta+\theta'=\frac{1}{2}\pi$; hence another way of stating the result is the following:—Draw a circle with A A′ (the mean points of the equal coils) as diameter: then if this circle cuts the common axis of the coils, the points of section are maxima, and the middle point is a minimum; but if the circle does not cut the axis, the middle point is a maximum, and the only one (see figure).

The law of decrease with distance is interesting: equation (44) shows us that when the two coils are close together and the coin is moved along the axis away from them, the effect which it is able to produce in the telephone varies pretty nearly as the inverse *sixth* power of the distance as soon as it has got a little way—a tremendously rapid rate of decrease.

It also shows that the effect varies directly with the fourth power of the diameter of the coin.

The Connexion between Induction-balance Effect and Conductivity.

20. The way in which the telephone-current depends on the conductivity $\frac{1}{\rho}$ of the coin is apparent in equation (40): it is evidently not simply proportional to it in general. But the self-induction λ of a circuit like that in a solid disk of metal must always be a very small quantity; and, except perhaps for highly conducting metals, it must probably be almost negligible compared with ρ. Assuming, then, that λ is infinitesimal, the expression simplifies, and ρ becomes a factor of one part of it. We get, in fact,

$$j = \frac{m\mu\text{E}(\text{S}-\text{R})}{\rho\text{R}(\text{L}r-l\text{S})}\left\{\frac{\text{S}}{\text{L}}e^{-\frac{\text{S}}{\text{L}}t} - \frac{r}{l}e^{-\frac{r}{l}t}\right\}$$
$$+\frac{m\mu\text{E}(\text{S}-\text{R})}{\text{R}(\text{L}-l)}e^{-\frac{\rho}{\lambda}t} \quad . \quad . \quad . \quad . \quad (45)$$

The term written by itself does not contain ρ except as an exponential; but it has a very great rate of variation depending directly on ρ, and it soon ceases to exist. The rest of the expression for the current is simply proportional to the conductivity, both as regards its own value and its rate of variation. What the precise meaning of all this is depends on what the loudness of the telephone-indication definitely depends on (§ 1); so for the present we will leave it in this state.

On Measuring with the Induction-balance.

21. Two methods have been employed both by Professor Hughes and by Mr Roberts, with the view of obtaining quantitative readings from the balance. The first consisted in estimating the loudness of the sound produced by the insertion of a coin, and then imitating it by an arrangement of primary and secondary coils called the "sonometer"—a key being used to transfer the connexions of the telephone quickly from one instrument to the other. Mr. Poynting has adapted a formula from Maxwell which suffices to graduate the sonometer when the secondary coil is some distance from either primary (Phil. Mag. January), though it is rather an unwieldy one.

But there must always be some objection to readings taken in this way, because of the difficulty of estimating precisely when two sounds have the same strength—especially if, as appears probable from § 18, they differ at all in quality, the sound produced by the coin in the balance having a tone mixed up with it which is shriller than any produced by the sonometer. (I do not know whether this has ever been noticed experimentally. One often notices a change of pitch in the telephone-rustle; but it appears to depend in some cases on the direction of the current, *i. e.* on whether it strengthens or weakens the magnet.)

If the sonometer is to be used, I would suggest that the second primary coil be done away with, and that the secondary coil be of small diameter. It may have as much wire on it as is wanted for long range; but the further it is away from the primary, compared with its own diameter, the better. The diameter of the primary does not much matter. The induction would now simply vary as the inverse cube of the distance from the mean circumference of the primary; hence, if the sonometer were graduated so as to give equal distances from this line, its readings would only have to be turned upside down and cubed to give comparative results. If both primary and secondary coils are small, the sonometer-arm may be graduated like an ordinary millimetre-scale.

If an absolute zero were wanted at any time, it might be obtained by an arrangement for rotating one of the coils till it was at right angles to the other.

22. Prof. Hughes's second method of measuring, that of the graduated zinc wedge, is a much better one, because it is a null method and gives true readings, though they are not easily interpretable.

A third method has, I believe, been tried, viz. a copper damper rotating above one pair of coils; and this seems also pretty good; but I think it might possibly be better to modify it by having a uniform disk (or ring) of high-conductivity copper capable of being moved along the axis of one pair of coils with its plane always parallel to theirs. The distance of the disk from the middle point of the coils should be read, and the distance apart of the coils should be constant. Given all particulars, I have no doubt that a mathematician could readily interpret results so obtained.

Another modification would be to use a coil of wire forming a circuit closed within itself instead of the copper disk. It would have just about the same effect as a solid disk of the same size; but the data connected with it might be more definite.

The Effect of Magnetic Bodies.

23. If a thin piece of iron wire be held in the balance, it powerfully disturbs it by a concentration of the lines of force in the iron, so that M (the mutual induction between the

battery- and telephone-circuits) is no longer zero. If the iron were instantaneously magnetizable, one ought to be able to balance its effect by moving the other pair of coils a little nearer together, so as to make M zero again. But as the magnetization of iron always takes time, this is not completely possible, because, though M may be made zero at any one instant, it will not remain so during the period of variation of the current. As regards pitch, the rustle produced by iron wire should agree with that produced by a shift of the coils, both being duller than that produced by a piece of copper. Of course if a *mass* of iron be inserted, the effect will be composite, secondary currents being induced in it which give rise to tertiary currents in the telephone-circuit. But this effect is feeble as compared with the magnetic effect.

In measuring the conductivity of a metal or of an electrolytic solution by means of the induction-balance, care must be taken that it is not magnetic; or erroneous results will be obtained. The conductivity effect and the magnetic effect will tend partially to destroy each other; for though the effect of a piece of copper cannot be completely balanced by introducing a scrap of iron into the same pair of coils, it can partially, just as it might partially by shifting the coils (§ 18).

It would be very interesting to observe an opposite effect with a diamagnetic body—say a large bundle of thin bismuth wires insulated from each other; but it would not be easy to make sure of the correctness of the observation.

Printed in Dunstable, United Kingdom

91220730R00020